D0272300
WITHDRAWN

Praise for DeLillo and *Zero K*

'Mysterious . . . Unexpectedly touching . . . [DeLillo offers] consolation simply by enacting so well the mystery and awe of the real world.'

—Joshua Ferris, *New York Times Book Review*

'DeLillo [has] almost Day-Glo powers as a writer.'

—Michiko Kakutani, *New York Times*

'Brilliant and astonishing . . . a masterpiece . . . manages to renew DeLillo's longstanding obsessions while also striking deeply and swiftly at the reader's emotions . . . The effect is transcendent.'

—Charles Finch, *Chicago Tribune*

'Daring . . . provocative . . . exquisite . . . captures the swelling fears of our age.'

—Ron Charles, *Washington Post*

THE SILENCE

Also by Don DeLillo

NOVELS

Americana
End Zone
Great Jones Street
Ratner's Star
Players
Running Dog
The Names
White Noise
Libra
Mao II
Underworld
The Body Artist
Cosmopolis
Falling Man
Point Omega
Zero K

STORIES

The Angel Esmeralda

PLAYS

The Day Room
Valparaiso
Love-Lies-Bleeding

THE SILENCE

a novel

Don DeLillo

PICADOR

First published 2020 by Scribner, an imprint of Simon & Schuster, Inc.
1230 Avenue of the Americas, New York, NY 10020

First published in the UK 2020 by Picador
an imprint of Pan Macmillan
The Smithson, 6 Briset Street, London EC1M 5NR
Associated companies throughout the world
www.panmacmillan.com

ISBN 978-1-5290-5709-6

5 7 9 8 6 4

A CIP catalogue record for this book is available from the British Library.

Printed and bound by CPI Group (UK) Ltd, Croydon, CR0 4YY

To Barbara Bennett

"I do not know with what weapons
World War III will be fought,
but World War IV will be fought
with sticks and stones."

Albert Einstein

PART ONE

-1-

Words, sentences, numbers, distance to destination.

The man touched the button and his seat moved from its upright position. He found himself staring up at the nearest of the small screens located just below the overhead bin, words and numbers changing with the progress of the flight. Altitude, air temperature, speed, time of arrival. He wanted to sleep but kept on looking.

Heure à Paris. Heure à London.

"Look," he said, and the woman nodded faintly but kept on writing in a little blue notebook.

He began to recite the words and numbers aloud because it made no sense, it had no effect, if he simply noted the changing details only to lose each one instantly in the twin drones of mind and aircraft.

"Okay. Altitude thirty-three thousand and two feet. Nice and precise," he said. "Température extérieure minus fifty-eight C."

He paused, waiting for her to say Celsius, but she looked at the notebook on the tray table in front of her and then thought a while before continuing to write.

"Okay. Time in New York twelve fifty-five. Doesn't say a.m. or p.m. Not that we have to be told."

Sleep was the point. He needed to sleep. But the words and numbers kept coming.

"Arrival time sixteen thirty-two. Speed four seventy-one m.p.h. Time to destination three thirty-four."

"I'm thinking back to the main course," she said. "I'm also thinking about the champagne with cranberry juice."

"But you didn't order it."

"Seemed pretentious. But I'm looking forward to the scones later in the flight."

She was talking and writing simultaneously.

"I like to pronounce the word properly," she said. "An abbreviated letter *o*. As in scot or trot. Or is it scone as in moan?"

He was watching her write. Was she writing what she was saying, what they were both saying?

She said, "Celsius. Cap *C*. It was someone's name. Can't recall his first name."

"Okay. What about vitesse. What does vitesse mean?"

"I'm thinking about Celsius and his work on the centigrade measurement."

"Then there's Fahrenheit."

"Him too."

"What does vitesse mean?"

"What?"

"Vitesse."

“Vitesse. Speed,” she said.

“Vitesse. Seven hundred forty-eight *k* per hour.”

His name was Jim Kripps. But for all the hours of this flight, his name was his seat number. This was the rooted procedure, his own, in accordance with the number stamped on his boarding pass.

“He was Swedish,” she said.

“Who?”

“Mr. Celsius.”

“Did you sneak a look at your phone?”

“You know how these things happen.”

“They come swimming out of deep memory. And when the man’s first name comes your way, I will begin to feel the pressure.”

“What pressure?”

“To produce Mr. Fahrenheit’s first name.”

She said, “Go back to your sky-high screen.”

“This flight. All the long flights. All the hours. Deeper than boredom.”

“Activate your tablet. Watch a movie.”

"I feel like talking. No headphone. We both feel like talking."

"No earbuds," she said. "Talk and write."

She was Jim's wife, dark-skinned, Tessa Berens, Caribbean-European-Asian origins, a poet whose work appeared often in literary journals. She also spent time, online, as an editor with an advisory group that answered questions from subscribers on subjects ranging from hearing loss to bodily equilibrium to dementia.

Here, in the air, much of what the couple said to each other seemed to be a function of some automated process, remarks generated by the nature of airline travel itself. None of the ramblings of people in rooms, in restaurants, where major motion is stilled by gravity, talk free-floating. All these hours over oceans or vast landmasses, sentences trimmed, sort of self-encased, passengers, pilots, cabin attendants, every word forgotten the moment the plane sets down on the tarmac and begins to taxi endlessly toward an unoccupied jetway.

He alone would remember some of it, he thought, middle of the night, in bed, images of sleeping people bundled into airline blankets, looking dead, the tall attendant asking if she could refill his wineglass, flight ending, seatbelt sign going off, the sense of release, passengers standing in the aisles, waiting, attendants at the exit, all their thank-yous and nodding heads, the million-mile smiles.

"Find a movie. Watch a movie."

"I'm too sleepy. Distance to destination, one thousand six hundred and one miles. Time in London eighteen o four. Speed four hundred sixty-five m.p.h. I'm reading whatever appears. Durée du vol three forty-five."

She said, "What time is the game?"

"Six-thirty kickoff."

"Do we get home in time?"

"Didn't I read it off the screen? Arrival time whatever whatever."

"We land in Newark, don't forget."

The game. In another life she might be interested. The flight. She wanted to be where she was going without this intermediate episode. Does anyone like long flights? She clearly was not anyone.

"Heure à Paris nineteen o eight," he said. "Heure à London eighteen o eight. Speed four hundred sixty-three m.p.h. We just lost two miles per hour."

"Okay I'll tell you what I'm writing. Simple. Some of the things we saw."

"In what language?"

"Elementary English. The cow jumped over the moon."

"We have pamphlets, booklets, entire volumes."

"I need to see it in my handwriting, perhaps twenty years from now, if I'm still alive, and find some missing element, something I don't see right now, if we're all still alive, twenty years, ten years."

"Filling time. There's also that."

"Filling time. Being boring. Living life."

"Okay. Température extérieure minus fifty-seven F," he said. "I'm doing my best to pronounce elementary French. Distance to destination one thousand five hundred seventy-eight miles. We should have contacted the car service."

"We'll jump in a taxi."

"All these people, a flight like this. They have cars waiting. The huge scramble at the exits. They know exactly where to go."

"They checked their baggage, most of them, some of them. We did not. Our advantage."

"Time in London eighteen eleven. Arrival time sixteen thirty-two. That was the last arrival time. Reassuring, I guess. Time in Paris nineteen eleven. Altitude thirty-three thousand and three feet. Durée du vol three sixteen."

Saying the words and numbers, speaking, detailing, allowed these indicators to live a while, officially noted, or voluntarily noted—the audible scan, he thought, of where and when.

She said, "Close your eyes."

"Okay. Speed four hundred seventy-six miles per hour. Time to destination."

She was right, let's not check our bags, we can squeeze them into the overhead. He watched the screen and thought about the game, briefly, forgetting who the Titans were playing.

Arrival time sixteen thirty. Température extérieure minus forty-seven C. Time in Paris twenty thirteen. Altitude thirty-four thousand and two feet. He liked the two feet. Definitely worth noting. Outside air temperature minus fifty-three F. Distance de parcours.

The Seahawks, of course.

Kripps was a tall man's name and he was tall, yes, but noncommittally so, and had no trouble meeting his need to be nondescript. He was not a proud head bobbing above a crowd but a hunched figure blessed by anonymity.

Then he thought back to the boarding process, all passengers seated finally, meal soon to appear, warm wet towels for the hands, toothbrush, toothpaste, socks, water bottle, pillow to go with the blanket.

Did he feel an element of shame in the presence of these features? They'd decided to fly business class despite the expense because the cramped space in tourist on a long flight was a challenge they wanted to avoid this one time.

Eye mask, face moisturizer, the cart with wines and liquors that an attendant pushes along the aisle now and then.

He watched the dangling screen and what he felt was the nudge of dumb indulgence. He thought of himself as strictly tourist. Planes, trains, restaurants. He never wanted to be well-dressed. It seemed the handiwork of a fraudulent second self. Man in the mirror, how impressed he is by the trimness of his image.

"Which was the rainy day?" she said.

"You're noting the rainy day in your book of memories. The rainy day, immortalized. The whole point of a holiday is to live it outstandingly. You've said this to me. To keep the high points in mind, the vivid moments and hours. The long walks, the great meals, the wine bars, the nightlife."

He wasn't listening to what he was saying because he knew it was stale air.

"Jardin du Luxembourg, Ile de la Cité, Notre-Dame, crippled but living. Centre Pompidou. I still have the ticket stub."

"I need to know the rainy day. It's a question of looking at the notes years from now and seeing the precision, the detail."

"You can't help yourself."

"I don't want to help myself," she said. "All I want to do is get home and look at a blank wall."

"Time to destination one hour twenty-six. I'll tell you what I can't remember. The name of this airline. Two weeks ago, starting out, different airline, no bilingual screen."

"But you're happy about the screen. You like your screen."

"It helps me hide from the noise."

Everything predetermined, a long flight, what we think and say, our immersion in a single sustained overtone, the engine roar, how we accept the need to accommodate it, keep it tolerable even if it isn't.

A seat that adapts to the passenger's wish for a massage.

"Speaking of remember. I remember now," she said.

"What?"

"Came out of nowhere. Anders."

"Anders."

"The first name of Mr. Celsius."

"Anders," he said.

"Anders Celsius."

She found this satisfying. Came out of nowhere. There is almost nothing left of nowhere. When a missing fact emerges without digital assistance, each person announces it to the other while looking off into a remote distance, the otherworld of what was known and lost.

“Children on this flight. Well-behaved,” he said.

“They know they’re not in economy. They sense their responsibility.”

She spoke and wrote simultaneously, head down.

“Okay. Altitude ten thousand three hundred sixty-four feet. Time in New York fifteen o two.”

“Except we’re going to Newark.”

“We don’t have to see every minute of the game.”

“I don’t.”

“I don’t,” he said.

“Of course you do.”

He decided to sleep for half an hour or until an attendant showed up with a snack before they landed. Tea and sweets. The plane began to bounce side to side. He knew that he was supposed to ignore this and that Tessa was supposed to shrug and say, Smooth ride up to now. The seatbelt sign flashed red. He tightened his seatbelt and looked at the screen while she went into a deeper crouch, her body nearly folding into her notebook. The bouncing became severe, altitude, air temperature, speed, he kept reading the screen but saying nothing. They were drowning in noise. A woman came staggering down the aisle, returning to the front row after a visit to the toilet, grabbing seatbacks for balance. Voices on the intercom, one of the pilots in French and then one of the attendants in English, and he thought that he might resume reading aloud from the screen but decided this would be a case of witless persistence in the midst of mental and physical distress. She was looking at him now, not writing just looking, and it occurred to him that he ought to move his seat to its upright position. She was already upright and she slid her food tray into the slot and put her notebook and pen in the seat pocket. A massive knocking somewhere below them. The screen went blank. Pilot speaking French, no English follow-up. Jim gripped the arms of his seat and then checked Tessa's seatbelt and retightened his. He imagined that every passenger was looking straight ahead into the six o'clock news, at home, on Channel 4, waiting for word of their crashed airliner.

"Are we afraid?" she said.

He let this question hover, thinking tea and sweets, tea and sweets.

-2-

Let the impulse dictate the logic.

This was the gambler's creed, his formal statement of belief.

They sat waiting in front of the superscreen TV. Diane Lucas and Max Stenner. The man had a history of big bets on sporting events and this was the final game of the football season, American football, two teams, eleven players each team, rectangular field one hundred yards long, goal lines and goal posts at either end, the national anthem sung by a semi-celebrity, six U.S. Air Force Thunderbirds streaking over the stadium.

Max was accustomed to being sedentary, attached to a surface, his armchair, sitting, watching, cursing silently when the field goal fails or the fumble occurs. The curse was visible in his slit eyes, right eye nearly shut, but depending on the game situation and the size of the wager, it might become a full-face profanity, a life regret, lips tight, chin quivering slightly, the wrinkle near the nose tending to lengthen. Not a single word, just this tension, and the right hand moving to the left forearm to scratch anthropoidally, primate style, fingers digging into flesh.

On this day, Super Bowl LVI in the year 2022, Diane was seated in the rocker five feet from Max, and between and behind them was her former student Martin, early thirties, bent slightly forward in a kitchen chair.

Commercials, station breaks, pregame babble.

Max, speaking over his shoulder, "The money is always there, the point spread, the bet itself. But consciously I recognize a split. Whatever happens on the field I have the point spread secured in mind but not the bet itself."

"Big dollars. But the actual amount," Diane

said, "is a number he keeps to himself. It is sacred territory. I am waiting for him to die first so he can tell me in his final breath how much money he has pissed away in the years of our something-or-other partnership."

"Ask her how many years."

The young man said nothing.

"Thirty-seven years," Diane said. "Not unhappily but in states of dire routine, two people so clutched together that the day is coming when each of us will forget the other's name."

A stream of commercials appeared and Diane looked at Max. Beer, whiskey, peanuts, soap and soda. She turned toward the young man.

She said, "Max doesn't stop watching. He becomes a consumer who had no intention of buying something. One hundred commercials in the next three or four hours."

"I watch them."

"He doesn't laugh or cry. But he watches."

Two other chairs, flanking the couple, ready for the latecomers.

Martin was always on time, neatly dressed, clean shaven, living alone in the Bronx where he taught high school physics and walked the streets unseen. It was a charter school, gifted students, and he was their semi-eccentric guide into the dense wonders of their subject.

“Halftime maybe I eat something,” Max said. “But I keep on watching.”

“He also listens.”

“I watch and I listen.”

“The sound down low.”

“Like it is now,” Max said.

“We can talk.”

“We talk, we listen, we eat, we drink, we watch.”

For the past year Diane has been telling the young man to return to earth. He barely occupied a chair, seemed only fitfully present, an original cliché, different from others, not a predictable or superficial figure but a man lost in his compulsive study of *Einstein's 1912 Manuscript on the Special Theory of Relativity*.

He tended to fall into a pale trance. Was this a sickness, a condition?

Onscreen an announcer and a former coach discussed the two quarterbacks. Max liked to complain about the way in which pro football has been reduced to two players, easier to deal with than the ever-shuffling units.

The opening kickoff was one commercial away. Max stood and rotated his upper body, this way and that, as far as it would go, feet firmly in place, and then looked straight ahead for about ten seconds. When he sat, Diane nodded as if allowing the proceedings onscreen to continue as planned.

The camera swept the crowd.

She said, "Imagine being there. Planted in a seat somewhere in the higher reaches of the stadium. What's the stadium called? Which corporation or product is it named after?"

She raised an arm, indicating a pause while she thought of a name for the stadium.

"The Benzedrex Nasal Decongestant Memorial Coliseum."

Max made a gesture of applause, hands not quite touching. He wanted to know where the others were, whether their flight was delayed, whether traffic was the problem, and will they bring something to eat and drink at halftime.

"We have plenty."

"We might need more. Five people. Long halftime. Singing, dancing, sex—what else?"

The teams trotted out to assume their respective positions. Kickoff team, receiving team.

Martin said, "What kept me completely engaged in the events on my TV screen was the World Cup. A global competition. Kick the ball, hit the ball with your head, do not touch the ball with your hands. Ancient traditions. Entire countries involved to the core. A shared religion. Team loses, players fall down on the field."

“Winning players also fall down on the field,” Diane said.

“People gathered in huge public squares in country after country, the World Cup, cheering, weeping.”

“Falling down in the street.”

“I watched once, briefly. Fake fucking injuries,” Max said. “And what kind of sport is it where you can’t use your hands? Can’t touch the ball with your hands unless you’re the goaltender. It’s like self-repression of the normal impulse. Here’s the ball. Grab it and run with it. This is normal. Grab it and throw it.”

“The World Cup,” Martin said again, close to a whisper. “I could not stop watching.”

Something happened then. The images onscreen began to shake. It was not ordinary visual distortion, it had depth, it formed abstract patterns that dissolved into a rhythmic pulse, a series of elementary units that seemed to thrust forward and then recede. Rectangles, triangles, squares.

They watched and listened. But there was nothing to listen to. Max picked up the remote control device from the floor in front of him and hit the volume button repeatedly but there was no audio.

Then the screen went blank. Max hit the power button. On, off, on. He and Diane checked their phones. Dead. She walked across the room to the house phone, the landline, a sentimental relic. No dial tone. Laptop, lifeless. She approached the computer in the next room and touched various elements but the screen stayed gray.

She returned to Max and stood behind him, hands on his shoulders, and she waited for him to clench his fists and start cursing.

He said calmly, "What is happening to my bet?"

He looked to Martin for an answer.

"Serious money. Where is my bet?"

Martin said, "It could be algorithmic governance. The Chinese. The Chinese watch the

Super Bowl. They play American football. The Beijing Barbarians. This is totally true. The joke is on us. They've initiated a selective internet apocalypse. They are watching, we are not."

Max shifted his gaze to Diane, who was seated again, looking at Martin. He was not a man who wisecracked about serious matters. Or were these the only matters he found funny?

Just then there was a snatch of dialogue coming from the blank screen. They tried to identify it. English, Russian, Mandarin, Cantonese? When it stopped, they waited for more. They looked, listened and waited.

"It is not earthly speech," Diane said. "It is extraterrestrial."

She wasn't sure whether she was the one who was joking now. She mentioned the military jets that had flown over the stadium ten or twelve minutes ago, or whenever it was.

Max said, "Happens every year. Our planes, a ritual flyover."

He repeated the last phrase and looked at Martin for confirmation of its eloquence.

Then he said, “An outdated ritual. We’ve gone beyond all comparisons between football and war. World Wars in Roman numerals, Super Bowls in Roman numerals. War is something else, happening somewhere else.”

“Hidden networks,” Martin said. “Changing by the minute, the microsecond, in ways beyond our imagining. Look at the blank screen. What is it hiding from us?”

Diane said, “Nobody is smarter than the Chinese except for Martin.”

Max was still looking at the young man.

“Say something smart,” he said.

“He quotes Einstein day and night. That’s pretty smart.”

“Okay, a footnote from the *1912 Manuscript*. ‘The beautiful and airy concepts of space and time.’ It’s not smart exactly but I keep repeating it.”

“In English or German?”

"Depends."

"Space and time," she said.

"Space and time. Spacetime."

"In class you quoted footnotes. You vanished into footnotes. Einstein, Heisenberg, Gödel. Relativity, uncertainty, incompleteness. I am foolishly trying to imagine all the rooms in all the cities where the game is being broadcast. All the people watching intently or sitting as we are, puzzled, abandoned by science, technology, common sense."

On an impulse she borrowed Martin's phone, thinking it might be more adaptable to the current circumstances. She looked at Max. She wanted to call their daughters, one in Boston, married, two kids, and the other somewhere in Europe on holiday. She hit buttons, shook the thing, stared into it, jabbed it with her thumbnail.

Nothing happened.

Martin said, "Somewhere in Chile."

She waited for more.

He said, "I'm sticking with Einstein no matter what the theorists have disclosed or predicted or imagined concerning gravitational waves, supersymmetries and so on. Einstein and black holes in space. He said it and then we saw it. Billions of times more massive than our sun. He said it many decades ago. His universe became ours. Black holes. The event horizon. The atomic clocks. Seeing the unseeable. North-central Chile. Did I say this?"

"You said everything."

"The Large Synoptic Survey Telescope."

"Somewhere in Chile. You said this."

Max faked a yawn.

"Let's return to here and now. What we have here is a communications screwup that affects this building and maybe this area and nowhere else and nobody else."

"So what do we do?"

"We talk to people who live in this building. Our so-called neighbors," he said.

He looked at her and then stood and shrugged and went out the door.

The two sat quietly for a moment. It occurred to her that she didn't know how to sit quietly with Martin.

"Something to eat."

"Maybe at halftime. If halftime ever comes."

"Einstein," she said. "The manuscript."

"Yes, the words and phrases that he crossed out. We can see him think."

"What else?"

"The nature of the handwritten text. The numbers, letters, expressions."

"What expressions?"

"'The force that the field exerts.' 'The theorem of the inertia of energy.'"

"What else?"

"'World point.' 'World line.'"

"What else?"

"'Weltpunkt.' 'Weltlinie.'"

"What else?"

"The way the facsimile pages become less pale but only briefly until the book nears the end."

"What else?"

"A slipcase, a hard cover, pages ten inches by fifteen inches. A big thing, I heft it, I turn the pages, I scan the pages."

"What else?" she said.

"This is Einstein, his handwriting, his formulas, his letters and numbers. The sheer physical beauty of the pages."

This was erotic in a way, this exchange. His responses were quick, his voice suggesting the eagerness of someone who has retained what truly matters.

She kept looking straight ahead at the blank screen.

"What else? What else?"

"Four words."

"What are they?"

"'Additional theorem of velocities.'"

"Say it again."

He said it again. She wanted to hear it one more time but she decided they ought to stop now. Teacher and student in a reverse coupling.

Martin Dekker. His full name, or most of it. She closed her eyes and said the name to herself. She said, Martin Dekker, will you live alone forever? The blank screen seemed a possible answer.

Then she turned and looked at him.

"So where is he? Where are the others?"

"Who are the others?"

"The two empty chairs. Old friends, more or less. Husband, wife. Returning from Paris, I think. Or Rome."

"Or north-central Chile."

"North-central Chile."

Max came back and went directly to the window across the room, looking down on empty Sunday streets. They talked about the doors he knocked on and the doors he bypassed. This became the main subject, doors as paneled structures worth describing, scratched, stained, recently painted. This floor, near neighbors, why get involved. One floor down, five doors, three responses, he said, holding up his hand, three fingers extended. Floor below that, four responses, two mentioned the game.

"We're waiting," she said.

"They saw and heard what we saw and heard. We stood in the hallway becoming neighbors for the first time. Men, women, nodding our heads."

"Did you introduce yourselves?"

"We nodded our heads."

"Okay. Important question. Is the elevator working?"

"I took the stairs."

"Okay. And did anyone have an idea about what is happening?"

"Something technical. Nobody blamed the Chinese. A systems failure. Also a sunspot. This was a serious response. A guy smoking a pipe. No, I did not tell him that smoking is not allowed in this building."

"Since you yourself. An occasional cigar," she said to Martin.

"A sunspot. A strong magnetic field. I stared at him."

"You gave him your death-penalty look."

"He said the experts will make adjustments."

Max stayed at the window, repeating this last remark in a whisper.

Diane waited for Martin to speak. She knew what she wanted him to say. But he didn't say it. So she attempted a playful version in the form of a question.

"Is this the casual embrace that marks the fall of world civilization?"

She forced a brief stab of laughter and waited for someone to say something.

-3-

Life can get so interesting that we forget to be afraid.

In the van, through the quiet streets, Jim waited for Tessa to look at him so they could trade looks.

There were others crammed into the vehicle, two flight attendants, a man talking to himself in French, a man talking to his phone, shaking it, cursing it. Others, moaning. Still others, quiet, trying to retrieve what had happened, who they were.

They were a wobbling mass of metal, glass and human life, down out of the sky.

Someone said, "We came down. I could not believe we were sort of floating."

Someone else said, "I don't know about floating. Maybe at first. But we hit hard."

"Did we miss the runway?"

"A crash landing. Flames," a woman said. "We were skidding and I looked out the window. Wing on fire."

Jim Kripps tried to remember what he saw. He tried to remember being afraid.

He had a cut on his forehead, a laceration, bloodless now. Tessa kept looking at it, almost wanted to touch it; maybe she thought this would help them remember. To touch, embrace, speak nonstop. Their phones were dead but this was no surprise. One of the passengers had a twisted arm, missing teeth. There were other injuries. The driver had told them that they were going to a clinic.

Tessa Berens. She knew her name. She had her passport, her money and her coat but no baggage or notebook, no sense of having gone through customs, no memory of fear. She was trying to bring things to mind more clearly. Jim was here and he was solid company, a man who worked as a claims adjuster for an insurance company.

Why was this so reassuring?

It was cold and dark but there was a jogger in the street, a woman in shorts and a T-shirt moving at a steady pace in the lane reserved for bicycles. They passed others here and there, hurrying, remote, just a few, no one sharing the barest glance.

"All we need is rain," Jim said, "and we'd know we were characters in a movie."

The cabin attendants were quiet, uniforms in slight disarray. Two or three questions directed their way from others in the van. Faint response, then nothing.

"We have to remember to keep telling ourselves that we're still alive," Tessa said, loud enough for the others to hear.

The man speaking French began to direct questions to the driver. Tessa tried to interpret for Jim.

The driver slowed down, keeping pace with the

running woman. He had no response to questions in any language. An elderly man said that he had to get to a toilet. But the driver did not increase the speed, clearly determined to stay aligned with the runner.

The woman just kept running, looking straight ahead.

-4-

How saints and angels haunt the empty churches at midnight, forgotten by the awed swarms of daytime tourists.

Max was back in his chair, cursing the situation. He kept looking at the blank screen. He kept saying *Jesus*, or *good Christ*, or *Jesus H. Christ.*

Diane sat at an angle now, able to watch both men. She told Max that this was a good time for him to prepare the halftime snack. It was possible, wasn't it, that reception would resume in a few minutes, the game in normal progress, and she added that she didn't believe a word of it.

Max went to the liquor cabinet instead of the kitchen and poured himself a glass of bourbon called Widow Jane, aged ten years in American oak.

On most occasions he would announce this to anyone in the room. *Aged ten years in American oak.* It was something he liked saying, a hint of irony in his voice.

This time he said nothing and did not offer to pour a glass for Martin. His wife drank wine but only with dinner, not with football.

He muttered the name *Jesus* several more times and sat looking at the screen, glass in hand, waiting.

Diane looked at Martin. She liked to do this. She pretended to study him. She thought of him as Young Martin, the title of a chapter in a book.

Then she said quietly, "Jesus of Nazareth."

Would Martin respond as she imagined he would?

"The radiant name," he said.

"We say this. You say it and I say it. What did Einstein say?"

"He said, 'I am a Jew but I am enthralled by the luminous figure of the Nazarene.'"

Max was staring into the blank screen. He looked and drank. Diane tried to keep her eyes on Martin. She knew that the name *Jesus of Nazareth* carried an intangible quality that drew him into its aura. He did not belong to a particular religion and did not feel reverence for any being of alleged supernatural power.

It was the name that gripped him. The beauty of the name. The name and place.

Max was leaning forward. He seemed to be trying to induce an image to appear on the screen through force of will.

Diane said, "Rome, Max, Rome. You remember this. Jesus in the churches and on the walls and ceilings of the palazzos. You remember better than I do. The one particular palazzo with tourists moving slowly room to room. Enormous paintings. The walls and ceilings. The one place in particular."

She looked at Martin. He was not a tiny cuddly childlike man. She thought of him as a mind trying to escape its commitment to the long slack body with flapping hands that seemed barely attached to his arms. She felt guilty for asking him to sit in a kitchen chair that didn't even have a cushioned seat.

"I tried to sneak us into a guided tour but Max wouldn't let me. He hated the idea of a guide," she said. "The paintings, the furniture, the statues in the long galleries. Arched ceilings with stunning murals. Totally, massively incredible."

She was looking into empty space now.

"Which palazzo?" she said to Max. "You remember. I do not."

Max sipped his drink, nodding slightly.

In one gallery tourists with headsets, motionless, lives suspended, looking up at the painted figure on the ceiling, angels, saints, Jesus in his garments, his raiment.

She spoke enthusiastically, head back, a momentary guide.

"How many years ago? Max."

He only nodded.

Martin said, "His raiment. I try to think of a rumpled garment embedded in the word."

"Others with audio guides hand-held, pressed to their ears. Voices in how many languages. I think of them even now, before I go to sleep, the still figures in the long galleries."

"Staring at the ceiling," Martin said.

"Max. When was it exactly? One year fades into the next. I'm getting older by the minute."

Max said, *"This team is ready to step out of the shadows and capture the moment."*

He seemed to be scrutinizing the blank screen.

The young man looked at the woman, the wife, the former professor, the friend, who found nothing, anywhere, to look at.

Max said, *"During this one blistering stretch, the offense has been pounding, pounding, pounding."*

She was reluctant to interrupt, to say something, anything, and finally she glanced over at Martin simply because it seemed essential to exchange a puzzled look with someone, anyone.

Max said, "*Avoids the sack, gets it away—intercepted!*"

It was time for another slug of bourbon and he paused and drank. His use of language was confident, she thought, emerging from a broadcast level deep in his unconscious mind, all these decades of indigenous discourse muddied up by the nature of the game, men hitting each other, men slamming each other into the turf.

"*Ground game, ground game, crowd chanting, stadium rocking.*"

Half sentences, bare words, repetitions. Diane wanted to think of it as a kind of plainsong, monophonic, ritualistic, but then told herself that this is pretentious nonsense.

Max speaking from deep in his throat, the voice of the crowd.

"*De-fense. De-fense. De-fense.*"

He got up, stretched, sat, drank.

"*Number seventy-seven, what's-his-name, looks*

bewildered, doesn't he? Penalty for spitting in opponent's face."

He said, "*These teams are evenly matched more or less. Punting from midfield. A barn burner of a game."*

Diane was beginning to be impressed.

He said, "*Coach of the offense. Murphy, Murray, Mumphrey, dialing up some innovations."*

He kept on talking, changing his tone, calm now, measured, persuasive.

"*Wireless the way you want it. Soothes and moisturizes. Gives you twice as much for the same low cost. Reduces the risk of heart-and-mind disease."*

Then, singing, "*Yes yes yes, never fails to bless bless bless."*

Diane was stunned. Is it the bourbon that's giving him this lilt, this flourish of football dialect and commercial jargon. Never happened before, not with bourbon, scotch, beer, marijuana.

She was enjoying this, at least she thought she was, based on how much longer he kept broadcasting.

Or is it the blank screen, is it a negative impulse that provoked his imagination, the sense that the game is happening somewhere in Deep Space outside the fragile reach of our current awareness, in some transrational warp that belongs to Martin's time frame, not ours.

Max said in a squeaky voice, "*Sometimes I wish I was human, man, woman, child, so I could taste this flavorful prune juice.*"

He said, "*Perpetual Postmortem Financing. Start your exclusive arrangements online.*"

Then, "*Play resumes, quarter two, hands, feet, knees, head, chest, crotch, hitting and getting hit. Super Bowl Fifty-Six. Our National Death Wish.*"

Diane whispered to Martin that there was no reason why they couldn't converse. Max had his game and he was beyond distraction.

The young man said quietly, "I've been taking a medication."

"Yes."

"The oral route."

"Yes. We all do this. A little white pill."

"There are side effects."

"A small pellet or tablet. White, pink, whatever."

"Could be constipation. Could be diarrhea."

"Yes," she whispered.

"Could be the feeling that others can hear your thoughts or control your behavior."

"I don't think I know about this."

"Irrational fear. Distrust of others. I can show you the insert," he said. "I carry it with me."

Max was scratching his forearm again, not with his fingers this time but with his knuckles.

He said, "*Field goal attempt from near midfield–fake, fake, fake!*"

The screen. Diane kept edging her head around to make sure it was blank. She could not understand why this was reassuring to her.

"Let's go down to the field," Max said. *"Esther, tell us what's happening."*

He raised his head now, phantom microphone in hand, and he spoke to a camera well above field level, his voice pitched to a higher tonal range.

"Here on the sidelines, this team exudes confidence despite the spate of injuries."

"The spate of injuries."

"That's right, Lester. I talked to the coordinator, offense, defense, whatever. He's happy as a pig in shit."

"Thank you, Esther. Now, back to the action."

It began to occur to Diane that Martin was speaking, although not necessarily to her.

"I look in the mirror and I don't know who I'm looking at," he said. "The face looking back at me doesn't seem to be mine. But then again why should it? Is the mirror a truly reflective surface? And is this the face that other people see? Or is it something or someone that I invent? Does the medication I'm taking release this second self?

I look at the face with interest. Interest and an element of confusion. Do other people experience this, ever? Our faces. And what do people see when they walk along the street and look at each other? Is it the same thing that I see? All our lives, all this looking. People looking. But seeing what?"

Max had stopped announcing. He was looking at Martin. They both were, husband and wife. The young man was peering into what is called the middle distance, staring carefully, in a measured way, and he was still talking.

"One escape is the movies. I tell my students. They sit and listen. Foreign-language films in black-and-white. Films in unfamiliar languages. A dead language, a subfamily, a dialect, an artificial language. Do not read subtitles. I tell them forthrightly. Avoid reading the printed translation of the spoken dialogue at the bottom of the screen. We want pure film, pure language. Indo-Iranian. Sino-Tibetan. People talking. They walk, talk, eat, drink. The stark power of black-and-white. The image, the optically formed duplicate. My students sit and listen. Smart young men and women. But they never seem to be looking at me."

"They're listening," Diane said, "and that's what matters."

Max was in the kitchen putting food on plates. She wanted to go for a walk, alone. Or she wanted Max to go for a walk and Martin to go home. Where are the others, Tessa and Jim and all the others, travelers, wanderers, pilgrims, people in houses and apartments and village hutments. Where are the cars and trucks, the traffic noises? Super Sunday. Is everyone at home or in darkened bars and social clubs, trying to watch the game? Think of the many millions of blank screens. Try to imagine the disabled phones.

What happens to people who live inside their phones?

Max returned to his bourbon. Diane realized that the young man was standing now, abandoning his customary slouch, head back, looking straight up.

She thought for a moment.

"The painted ceilings. Rome," she said. "The tourists looking up."

"Standing absolutely still."

"Saints and angels. Jesus of Nazareth."

"The luminous figure. The Nazarene. Einstein," he said.

-5-

Lost systems in the crux of everyday life.

The clinic was a sprawl of halls and rooms at street level and Jim and Tessa walked past doors, exit signs, blinking red lights, posted notices lettered by hand. Staff members hurried past wearing street clothes under their flapping tunics.

Other people from the van entered rooms or formed lines or stood around talking. A few had remained in the vehicle, destination unknown.

There was a woman crouched down on a stool in a cramped space, a cubbyhole.

"The administrator," Tessa said. "The functionary."

They joined a long line of people waiting to see the woman. The hall lights kept dimming.

After a time Jim said, "Why are we standing here?"

"You have a wound."

"A wound. On my head. I forgot."

"You forgot. Let me have a look," Tessa said. "A gash. Shapeless. When we crash-landed and undid our seatbelts and jumped up to get the hell out of there, I saw that you were bleeding."

"Hit my head on the window."

"Let's be patient and wait on line and then see what the official on the stool has to say."

"But first."

"But first," she said.

They left the line and eventually found a vacant toilet. In that scant space he eased her against a bare wall and she opened her coat and unbuckled his belt and pulled down his pants and shorts and asked him if his head hurt and he responded by undressing her slowly and carefully and they talked about what they were doing, how, where, when, suggesting, advising, trying not to laugh, her body slowly lowering along the wall and he buckled his knees to maintain distance and rhythm.

Someone was knocking on the door, then speaking into it. *Show some consideration.* Another voice then, accented. Tessa whispered a list of nationalities as they completed the act and crudely wiped each other with tissues from the dispenser adjacent to the mirror.

They finished dressing and looked at each other for a long moment. This look summed up the day and their survival and the depth of their connection. The state of things, the world outside, this would require another kind of look whenever it became appropriate.

Then they went out the door and down the hall. The line was much shorter now and they decided to take their place and wait.

"I guess we can walk to their building from here if that's the only way to get there."

"They're our friends. They'll feed us."

“They’ll listen to our story.”

“They’ll tell us what they know.”

“The Super Bowl. Where is it happening?”

“Somewhere temperate, in sunlight and shadow,” he said, “before shouting thousands.”

The woman in the cubbyhole looked up at them, another set of faces and bodies, all day, people standing, talking, listening, waiting for instructions concerning where to go, who to see, which hall, which door, and she nodded as if knowing in advance who they were and what they wanted.

She seemed glued to the stool she was sitting on.

“Our plane, we experienced a crash landing,” Tessa said. “He suffered an injury.”

Jim towered above the woman and he leaned over and pointed to the wound, feeling like a schoolboy injured at playtime.

“I have nothing to do with actual human bodies. No look, no touch. I will send you to an examining room,” she said, “where a trained individual will either treat you or send you to someone else somewhere else. Everyone I've seen today has a story. You two are the plane crash. Others are the abandoned subway, the stalled elevators, then the empty office buildings, the barricaded storefronts. I tell them that we are here for injured people. I am not here to dispense advice concerning the current situation. What *is* the current situation?"

She pointed to the blank screen in the panel of devices set into the wall in front of her. She was middle-aged, dressed in high boots, sturdy jeans, a heavy sweater, with rings on three fingers.

"I can tell you this. Whatever is going on, it has crushed our technology. The word itself seems outdated to me, lost in space. Where is the leap of authority to our secure devices, our encryption capacities, our tweets, trolls and bots. Is everything in the datasphere subject to distortion and theft? And do we simply have to sit here and mourn our fate?"

Jim was still stooped over, displaying his

wound. The woman leaned forward and twisted her head to look up at him.

She said, "Why am I telling you this? Because your plane crashed, more or less, and you are eager to hear what is going on. And because I'm still a talky little kid when the circumstances warrant."

Tessa said, "We're here to listen."

The overhead lights blinked and dimmed and then went out. There was instant silence throughout the clinic. Everyone waiting. A sense also of fear-in-waiting because it wasn't clear yet what this might mean, how radical, how permanent an aberration in what was already a drastic shift of events.

The woman spoke first, in a whisper, telling them where she was born and raised, names of parents and grandparents, sisters and brothers, schools, clinics, hospitals, her voice suggesting an intimate calm tinged with hysteria.

They waited.

She resumed with her first marriage, first

cellphone, divorce, travel, French boyfriend, riots in the streets.

They waited some more.

"No e-mail," she said, leaning back, palms up. "More or less unthinkable. What do we do? Who do we blame?"

Gestures barely visible.

"E-mail-less. Try to imagine it. Say it. Hear how it sounds. E-mail-less."

Her head bouncing slightly with each spoken syllable. Someone with a flashlight was standing in the doorway, training the beam on each of them, once and then again, before leaving without a word.

A brief pause and then the woman resumed speaking in the dark, her whisper more intense now.

"The more advanced, the more vulnerable. Our systems of surveillance, our facial recognition devices, our imagery resolution. How do we know who we are? We know it's getting cold in here. What happens when we have to leave? No light, no heat. Going home, living where I live, above a restaurant called Truth and Beauty, if the subways and buses are not running, if the taxis are gone, elevator in the building immobilized, and if, and if, and if. I love my cubicle but I don't want to die here."

She was quiet for a time. When the lights returned, dimly, Jim was standing straight up, expressionless. A tall white android.

The woman spoke in a normal voice now.

"Okay I see the wound and I can say without hesitation that you need to go down the hall, third room on the left."

She pointed in that direction and then put on a pair of wool gloves and pointed again, authoritatively.

"And when you're finished there, then what?"

"We're going to see friends," Tessa said. "As originally planned."

"How will you get there?"

"Walk."

"Then what?" the woman said.

"Then what?" Jim said.

They waited for Tessa to add her voice to the elemental dilemma but she simply shrugged.

In a room down the hall a young man in an oversized tunic and a baseball cap stood on his toes to brush a medication on Jim's wound and then bandage it securely. Jim started to shake his hand and then changed his mind and they left.

Out in the street they talked about the woman they'd seen jogging when they were in the van. It would have been encouraging to see her again. The wind was fierce and they walked quickly, heads down. The only person to be seen was a hobbling man pushing a battered cart that probably contained everything he owned. He paused to wave at them and then stepped away from the cart and took a few long strides, body bent, to imitate their movements. They waved back and kept on going. At a major intersection the digital traffic warden was stilled, one levered arm raised slightly.

There was nothing for them to do but keep on walking.

-6-

Counting down by sevens in the future that takes shape too soon.

There were six candles placed around the living room and Diane had just put a match to the last of them.

She said, “Is this a situation where we have to think about what we’re going to say before we say it?”

“The semi-darkness. It’s somewhere in the mass mind,” Martin said. “The pause, the sense of having experienced this before. Some kind of natural breakdown or foreign intrusion. A cautionary sense that we inherit from our grandparents or great-grandparents or back beyond. People in the grip of serious threat.”

“Is that who we are?”

“I’m talking too much,” he said. “I’m grinding out theories and speculations.”

The young man was standing at the window and Diane wondered if he planned to head home to the Bronx. She imagined that he might have to walk all the way, up through East Harlem to one of the bridges. Were pedestrians allowed to cross or were the bridges for cars and buses only? Was anything operating normally out there?

The thought softened her, made her think that she might offer to accommodate him for the night. The sofa, a blanket, not so complicated.

Stove dead, refrigerator dead. Heat beginning to fade into the walls. Max Stenner was in his chair, eyes on the blank screen. It seemed to be his turn to speak. She sensed it, nodded and waited.

He said, “Let’s eat now. Or the food will go hard or soft or warm or cold or whatever.”

They thought about this. But nobody moved in the direction of the kitchen.

Then Martin said, “Football.”

A reminder of how the long afternoon had started. He made a gesture, strange for such an individual, the action in slow motion of a player throwing a football, body poised, left arm thrust forward, providing balance, right arm set back, hand gripping football.

Here was Martin Dekker and there was Diane Lucas standing across the room, puzzled by the apparition.

He seemed lost in the pose but returned eventually to a natural stance. Max was back to his blank screen. The pauses were turning into silences and beginning to feel like the wrong kind of normal. Diane waited for her husband to pour more whiskey but he showed no interest, at least for now. Everything that was simple and declarative, where did it go?

Martin said, “Are we living in a makeshift reality? Have I already said this? A future that isn't supposed to take form just yet?”

"A power station failed. That's all," she said. "Consider the situation in those terms. A facility along the Hudson River."

"Artificial intelligence that betrays who we are and how we live and think."

"Lights back on, heat back on, our collective mind back where it was, more or less, in a day or two."

"The artificial future. The neural interface."

They seemed determined not to look at each other.

Martin, speaking to no one in particular, raised the subject of his students. Global origins, assorted accents, all smart, specially selected for his course, ready for anything he might say, whatever assignment, whatever proposal he might advance concerning areas of study beyond physics. He'd recited names to them. Thaumatology, ontology, eschatology, epistemology. He could not stop himself. Metaphysics, phenomenology, transcendentalism. He paused and thought and kept going. Teleology, etiology, ontogeny, phylogeny.

They looked, they listened, they sniffed the stale air. This is why they were there, all of them, students and teacher.

"And one of the students recited a dream he'd had. It was a dream of words, not images. Two words. He woke up with those words and just stared into space. *Umbrella'd ambuscade. Umbrella* with an apostrophe *d*. And *ambuscade*. He had to look up the latter word. How could he dream of a word he'd never encountered? *Ambuscade*. Ambush. But it was *umbrella* with an apostrophe *d* that seemed a true mystery. And the two words joined. *Umbrella'd ambuscade*."

He waited for a time.

"All this in the Bronx," he said finally, making Diane smile. "There I stood listening to the young men and women discuss the matter, the students, my students, and I wondered, myself, what to make of the term. Ten men with umbrellas? Preparing an attack? And the student whose dream it was, he was looking at me as if I were responsible for what happened in his sleep. All my fault. Apostrophe *d*."

There was a knock on the door. It sounded weary, elevators not working, people having to climb eight flights. Diane was standing right there but paused before reaching for the doorknob.

"I was hoping it was you."

"It's us, barely," Jim Kripps said.

They took off their coats and tossed them on the sofa and Diane gestured to Martin and spoke his name and there were handshakes and half embraces and Max standing with one clenched fist raised in a gesture of greeting. He saw the bandage on Jim's forehead and threw a few counterfeit punches.

When everyone was seated, here, there, the newcomers spoke of the flight and the events that followed and the spectacle of the midtown streets, the grid system, all emptied out.

"In darkness."

"No street lights, store lights, high-rise buildings, skyscrapers, all windows everywhere."

"Dark."

"Quarter-moon up there somewhere."

"And you're back from Rome."

"We're back from Paris," Tessa said.

Diane thought she was beautiful, mixed parentage, her poetry obscure, intimate, impressive.

The couple lived on the Upper West Side, which would have meant a walk through Central Park in total darkness and then a longer walk uptown.

The conversation became labored after a while, shadowed in disquiet. Jim spoke looking down between his feet and Diane waved her arms indicating events taking place somewhere beyond their shallow grasp.

"Food. Time to eat something," she said. "But first I'm curious about the food they served on your flight. I know I'm babbling. But I ask people this question and they never remember. Ask about the last restaurant meal even if it was a week ago and they can tell me. No problem. Name of restaurant, name of main course, type of wine, country of origin.

But food on planes. First class, business class, economy, none of it matters. People do not remember what they ate."

"Spinach-and-cheese tortellini," Tessa said.

No one spoke for a moment.

Then Diane said, "Our food. Here and now. Football food."

Martin went with her to the kitchen. The others waited quietly in candlelight. Soon Tessa started counting down slowly by sevens from two hundred and three to zero, deadpan, changing languages along the way, and eventually the food arrived, prepared earlier by Max, and all five individuals sat and ate. The kitchen chair, the rocking chair, the armchair, a side chair, a folding chair. None of the guests offered to go home after the meal even when Jim and Tessa got their coats off the sofa and put them back on, simply needing to get warmer. Martin closed his eyes as he chewed his food.

Was each a mystery to the others, however close their involvement, each individual so naturally encased that he or she escaped a final determination, a fixed appraisal by the others in the room?

Max looked at the screen as he ate and when he was finished eating he put the plate down and kept on looking. He took the bottle of bourbon off the floor and the glass with it and poured himself a drink. He put the bottle down and held the glass in both hands.

Then he stared into the blank screen.

PART TWO

It is clear by now that the launch codes are being manipulated remotely by unknown groups or agencies. All nuclear weapons, worldwide, have become dysfunctional. Missiles are not soaring over oceans, bombs are not being dropped from supersonic aircraft.

But the war rolls on and the terms accumulate.

Cyberattacks, digital intrusions, biological aggressions. Anthrax, smallpox, pathogens. The dead and disabled. Starvation, plague and what else?

Power grids collapsing. Our personal perceptions sinking into quantum dominance.

Are the oceans rising rapidly? Is the air getting warmer, hour by hour, minute by minute?

Do people experience memories of earlier conflicts, the spread of terrorism, the shaky video of someone approaching an embassy, a bomb vest strapped to his chest? Pray and die. War that we can see and feel.

Is there a shred of nostalgia in these recollections?

People begin to appear in the streets, warily at first and then in a spirit of release, walking, looking, wondering, women and men, an incidental cluster of adolescents, all escorting each other through the mass insomnia of this inconceivable time.

And isn't it strange that certain individuals have seemed to accept the shutdown, the burnout? Is this something that they've always longed for, subliminally, subatomically? Some people, always some, a minuscule number among the human inhabitants of planet Earth, third planet from the sun, the realm of mortal existence.

"Nobody wants to call it World War III but this is what it is," Martin says.

Seemingly all screens have emptied out, everywhere. What remains for us to see, hear, feel? Do a select number of people have a form of phone implanted in their bodies? A serious question, the young man says. Is this a protection against the global silence that marks our hours, minutes and seconds? Who are such people? How do they access the subcutaneous calls? Is there a body-code, a sort of second heartbeat that conveys a local warning?

It is well past midnight and he is still talking and Diane is still listening and the friends are still here, Jim Kripps and Tessa Berens, with Max crunched in his chair.

Dark energy, phantom waves, hack and counterhack.

Mass surveillance software that makes its own decisions, overruling itself at times.

Satellite tracking data.

Targets in space that remain in space.

All in the living room, all in coats, three wearing gloves as well, four of them apparently listening to Martin, the one standing person, gesturing freely as he speaks.

The way in which time has seemingly jumped forward. Did something happen at midnight to intensify the disruption? And the way in which Martin's voice is beginning to change.

Bioweapons and the countries that possess them.

He recites a long list that is interrupted by a coughing fit. The others look away. He wipes his mouth with the back of his hand, then inspects the hand and continues talking.

Certain countries. Once rabid proponents of nuclear arms, now speaking the language of living weaponry.

Germs, genes, spores, powders.

Diane begins to understand that he is using an accent. Not just a voice speaking in a manner not his own but a voice meant to belong to a particular individual. This is Martin's version of Albert Einstein speaking English.

She is not sure that what he is saying is pure fiction. Something about him, his tone of voice, adopted accent, a sense that he has access to world events, whatever that means, however he is able to allow censored news to reach him. He said it himself, people with phones implanted in their bodies.

She understands that this is foolish, all of it. She also knows that there is something in her former student's essential nature that makes such speculation possible.

She is babbling again but only to herself this time.

She decides to say nothing to the others about the accent that Martin is using. He speaks more softly now, hands caressing the words.

Wave structure, metric tensor, covariant qualities.

It may be too complicated to bring Einstein into the room. And she doesn't know whether these are terms out of the *1912 Manuscript*, Martin's bible, his playbook, or simply noises floating in the air, the language of World War III.

He sounds either brilliant or unbalanced, Martin does, not Einstein, as he recites the names of those scientists attending a conference in Brussels in 1927, twenty-eight men and one woman, Marie Curie, Madame Curie, name after name, with Einstein referring to himself in Martin's voice as Albert Einstein–seated-front-and-center.

And now he swings from accented English into living German. Diane tries to follow what he is saying but quickly loses all sense of it. There is no hint of parody or self-parody. It is all in

Martin's mind as he stands alone at the mirror in his apartment except that he is not there, he is here, thinking aloud, drawing inward, shaking his head.

Einstein's parents were Pauline and Hermann.

She understands this simple sentence but does not try to keep listening. She wants him to stop and will tell him so. He stands up straight, speaking earnestly either as himself or as Einstein, and does it matter?

Max stands and stretches. Max Stenner. Max. This is all it takes to silence the young man.

"We're being zombified," Max says. "We're being bird-brained."

He walks toward the front door, talking to them over his shoulder.

"I'm done with all this. Sunday or is it Monday? February whatever. It's my expiration date."

Nobody knows what he means by this.

He zippers his jacket and leaves and Diane

thinks of him walking down the stairs, one step, then another. Her mind is operating in slow motion now. She almost feels obliged to sit in front of the TV set on his behalf, waiting for something to splash onto the screen.

Martin resumes speaking for a time, back to English, unaccented.

Internet arms race, wireless signals, countersurveillance.

"Data breaches," he says. "Cryptocurrencies."

He speaks this last term looking directly at Diane.

Cryptocurrencies.

She builds the word in her mind, unhyphenated.

They are looking at each other now.

She says, "Cryptocurrencies."

She doesn't have to ask him what this means.

He says, "Money running wild. Not a new development. No government standard. Financial mayhem."

"And it is happening when?"

"Now," he says. "Has been happening. Will continue to happen."

"Cryptocurrencies."

"Now."

"Crypto," she says, pausing, keeping her eyes on Martin. "Currencies."

Somewhere within all those syllables, something secret, covert, intimate.

Then Tessa speaks.

She says, “What if?”

This results in a long pause, a shift in mood. They wait for more.

“What if all this is some kind of living breathing fantasy?”

“Made more or less real,” Jim says.

“What if we are not what we think we are? What if the world we know is being completely rearranged as we stand and watch or sit and talk?”

She raises a hand and lets the fingers flap up and down in a gesture of everyday babble.

“Has time leaped forward, as our young man says, or has it collapsed? And will people in the streets become flash mobs, running wild, breaking and entering, everywhere, planet-wide, rejecting the past, completely unmoored from all the habits and patterns?”

No one moves toward the window to look.

"What comes next?" Tessa says. "It was always at the edges of our perception. Power out, technology slipping away, one aspect, then another. We've seen it happening repeatedly, this country and elsewhere, storms and wildfires and evacuations, typhoons, tornadoes, drought, dense fog, foul air. Landslides, tsunamis, disappearing rivers, houses collapsing, entire buildings crumbling, skies blotted out by pollution. I'm sorry and I'll try to shut up. But remaining fresh in every memory, virus, plague, the march through airport terminals, the face masks, the city streets emptied out."

Tessa notes the silence that attends her pauses.

"From the one blank screen in this apartment to the situation that surrounds us. What is happening? Who is doing this to us? Have our minds been digitally remastered? Are we an experiment that happens to be falling apart, a scheme set in motion by forces outside our reckoning? This is not the first time these questions have been asked. Scientists have said things, written things, physicists, philosophers."

In the second silence all heads turn toward Martin.

He speaks of satellites in orbit that are able to see everything. The street where we live, the building we work in, the socks we are wearing. A rain of asteroids. The sky thick with them. Could happen anytime. Asteroids that become meteorites as they approach a planet. Entire exoplanets blown away.

Why not us. Why not now.

"All we have to do is consider our situation," he says. "Whatever is out there, we are still people, the human slivers of a civilization."

He lets the phrase linger. The human slivers.

Tessa begins to separate herself. She seeps away to the sound of the young man's voice. She thinks into herself. She sees herself. She is different from these people. She imagines taking off her clothes, nonerotically, to show them who she is.

Be serious. Be here. Or what about somewhere nearby, the bedroom. They've had near-death, they've had sex, they need sleep, and she looks at Jim, leaning her head just slightly toward the hallway.

He asks Diane about the bedroom. A long flight, a long day, a brief sleep would be nice.

She watches them walk down the hall. In the dwindling spirit of this astonishing time, she isn't surprised. Sleep, obviously, understandably, after what they've experienced. She tries to remember whether she made the bed this morning, cleaned the room. Max sometimes cleans; he cleans and then inspects, scrupulously.

There is only one bedroom, one bed, but let it belong to Jim Kripps and Tessa whatever-her-last-name-is. They'll be headed home at first light.

Martin is speaking again.

"The drone wars. Never mind country of origin. The drones have become autonomous."

He begins to notice that he and Diane are the only ones left in the living room.

"Drones above us now. Flinging warnings at each other. Their weapon being a form of the language isolate. A language known only to drones."

How did this happen, five people down to two. The man remains standing and they look at each other. The woman realizes that she is still in the thrall of cryptocurrencies.

She says the word and waits for him to respond.

Finally he says, "Cryptocurrencies, microplastics. The dangers at every level. Eat, drink, invest. Breathe, inhale, draw oxygen into the lungs. Walk, run, stand. And now in the purest snow from the alpine wilderness, from the arctic wasteland."

"What?"

"Plastics, microplastics. In our air, our water, our food."

She had hoped to hear something libidinal, arousing. She understands that he has something more to say and she looks and waits.

He says, "Greenland is disappearing."

She gets to her feet and faces him.

She says, "Martin Dekker. You know what we want, don't you?"

They could sidestep their way into the kitchen and she could stand with her back pressed to the two vertical bars of the refrigerator door and they could do it quickly, forgettably, in the spirit of the onward moment.

He unbuckles his belt and drops his pants. He stands there, stricken, in his checkered shorts, looking taller than ever. She tells him to say something in German and when he does, a substantial statement recited quickly, she asks for a translation.

He says, "Capitalism is an economic system in which the means of production and distribution are privately or corporately owned, and development is proportionate to the accumulation and reinvestment of profits gained in a free market."

She nods, half-smiling, and motions for him to lift his trousers and buckle his belt. She finds it satisfying to mimic his belt-buckling. She understands that sex with her former student may be a sleazy little tremor in her mind but is nowhere present in her body.

She expects him to walk out the door and hates to think of him trying to get home in whatever circumstances now prevail. Instead he takes three long strides to the nearest chair and sits there, looking into space.

In the bedroom Tessa thinks about going home, being home, the place, finally, where they don't see each other, walk past each other, say *what* when the other speaks, aware only of a familiar shape making noise somewhere nearby.

Jim is nearby now, next to her on the bed, asleep, his body shaking slightly.

There is a poem she wants to work on, tomorrow, next day, when she is fully awake, at her desk at home, the first line bouncing around in her brain for a while.

In a tumbling void.

She will see the line when she closes her eyes and concentrates. See the letters set against a dark background and then slowly open her eyes to whatever is in front of her, dominant objects only inches high, a paperweight, a photograph, a toy taxicab.

Jim is awake now. He takes a long time to build to a spacious yawn. Tessa says something in a language that he does not recognize until he realizes it is simply fake, a dead language, a dialect, an idiolect (whatever that is) or something else completely.

"Home," he says finally. "Where is that?"

Max is making his way through the crowded streets when he grudgingly recalls something the young man said and wonders if what he is seeing here and now is an aspect of Martin Dekker's mind repositioned in three dimensions.

Is it like this in other cities, people on a rampage, nowhere to go? Do crowds in a Canadian city spill down to join crowds here? Is Europe one impossible crowd? What time is it in Europe? Are the public squares swarming with people, tens of thousands, and all of Asia and Africa and elsewhere?

Names of countries keep rolling through his mind and people are trying to talk to him and to each other and he thinks of his daughter with two kids and a husband in Boston and the other daughter traveling somewhere and for one strange and compressed and claustrophobic moment he forgets their names.

He stands against a wall and watches.

In other times, more or less ordinary, there are always people staring into their phones, morning, noon, night, middle of the sidewalk, oblivious to everyone hurrying past, engrossed, mesmerized, consumed by the device, or walking toward him and then veering away, but they can't do it now, all the digital addicts, phones shut down, everything down down down.

He tells himself that it's time to head home and that he will have to muscle his way through the crowd, people hunched against the cold, a thousand faces every minute, people wrestling, throwing punches, a small riot here and there, curses rising into the air. He stands for a few seconds longer, flexing shoulders in preparation, and decides that when he gets to his building he will count the steps as he climbs to the apartment. He has done this before but not for many decades and begins to wonder what the point is.

Then he walks into the streaming mass.

Diane, back home, where else, is trying to suppress a series of little squeaky belches.

She says, “Somewhere in Chile.”

This seems to mean something but she can’t remember what it is. She looks at Martin and then at the other two, returning from the bedroom. The man is yawning and the woman is almost fully dressed, wearing low-cut socks but no shoes. Diane mutters a few foul words, mocking herself for yielding to the spirit of the moment and allowing the bedroom to be used by sex-frenzied guests.

Or maybe they were just resting. This is what they’d said, this is what she’d originally believed.

Martin says, “The Cerro-Pachón Ridge in north-central Chile.”

“What’s that?” Jim says.

“The Large Synoptic Survey Telescope.”

He goes on to explain the matter and then Max

walks in and unzips his jacket. They wait for him to say something. He takes off the jacket and tosses it on the floor next to the remote control device, his bottle of bourbon and his empty glass. He fills the glass and drinks, shaking his head at the bracing shock of raw whiskey.

What is happening in the streets? What is out there? Who is out there?

He says, “You don’t want to know.”

Then he raises his glass.

“Widow Jane,” he says. “Aged ten years in American oak. Did I say this before?”

He drinks and then leans forward and to his left, looking at Tessa’s feet.

“What happened to your shoes?”

“They walked off without me,” she says.

Everybody feels better now.

Martin is not finished. He says, "The onward moments, the flowing moments. People have to keep telling themselves that they're still alive."

Jim Kripps listens to himself breathe. Then he touches the bandage on his forehead, just checking, confirming that it's still there.

Two of the others are barely awake, Tessa and Max. Diane understands that she is here to listen to her former student as he had once listened to her.

"When we're finished with all this, it may be time for me to embrace a free death. *Freitod*," he says. "But am I serious about this or simply begging attention? And the situation we are in. Shouldn't I be home, alone, in my room? Isn't this what the circumstances warrant? No word from anyone, anywhere. Time to sit and be still."

He touches the edges of his chair, confirming the fact that he is seated.

"Or am I being a little too self-important?" he says, slowly, drawing out the question, hands stiffening, gaze seeming to recede as he begins to enter the trancelike state that she has seen before and that she thinks of as metaphysical.

"All my life I've been waiting for this without knowing it," he says.

Diane Lucas decides to say something, although she has no idea what might come spouting out.

"Staring into space. Losing track of time. Going to bed. Getting out of bed. Months and years and decades of teaching. Students tend to listen. All those different backgrounds. The faces dark, light, medium. What is happening in the public squares across Europe, the places where I've walked and looked and listened? I feel so simpleminded. A college professor who quit too soon. A would-be inspiration to my students, one of whom sits next to me here and now. The end-of-the-world movie. People stranded in a room. But we're not stranded. We can leave anytime. I try to imagine the vast sense of confusion out there. My husband does not want to describe what he has seen but I am guessing bedlam in the streets and why am I so reluctant to get up and walk to the window and simply look? But didn't this have to happen? Isn't that what some of us are thinking? We were headed in this direction. No more wonder, no more curiosity. Totally impaired orientation. Too much of everything from too narrow a source code. And am I saying all this because it's way past midnight and I haven't slept and have barely eaten and the people here with me are barely listening to what I'm saying? Tell me I'm wrong, someone, but of course no one speaks. I want to resume teaching and return to my classroom and speak to my students about the principles of physics. The physics of this, the physics of that. The physics of time. Absolute time. Time's arrow. Time and space. Before I shut up I will quote a stray line from *Finnegans Wake*, a book I've been reading on and off, here and there, for what seems like forever. The line has stayed secure in the proper mind slot, the word *preserve*. *Ere the sockson locked at the dure*. Just one more thing to say. To myself this time. Shut up, Diane."

Jim Kripps is stooped low in his chair, looking down, speaking into the carpet, long hands dangling.

"So there we were, seated, half-asleep, waiting for a snack before we landed. Then the trouble began. Our plane was jumping around and making loud banging noises. I don't think we were anywhere near the airport, the landing strip. Bang bang bang. I looked out the window, seeing nothing, waiting for some reassurance from the pilot. Here is Tessa seated next to me, just as she was then. I don't think I looked at her because I didn't want to see the look on her face. The plane was wobbling badly. Then voices on the intercom, totally unreassuring. This is how it begins, this is how it feels, all those many thousands of passengers before us who have experienced this and then were silenced forever. Did this occur to me, those many thousands, or am I making this up as I ramble along? It seems a dozen years ago but this was today, more or less, just hours ago, how many hours, pilot speaking French, our seatbelts, our snack before landing, where was our fucking snack. Tessa speaks French. Did she translate for me? I don't think so and she was probably doing me a favor. Sorry to be going on like this and then the crash landing, a huge sort of rocketing noise and the impact that felt like God's own voice, forgive me, and my head hit the window, I was tossed sideways into the window, someone shouting *fire*, was a wing on fire, and I felt the blood running into my eye and reached for Tessa's hand, she's here, she's saying something, and someone across the aisle half-choking, half-shouting. *No no no.* Well, anyway, to make a short story even shorter, we came down hard and skipped and skidded for a while and of course I had no way until later to connect this event with the total collapse of all systems and my hand was on Tessa's wrist and she was looking at the blood on my face. This is the first chance I've had to really think about it, to remember it. Before this, the van, the clinic, the woman talking talking talking, the man in the baseball cap bandaging my head. Into the street. A young woman jogging."

Max Stenner is trying to look bored. He sits in his chair, the armchair, eyes barely open.

"The stairs. Coming back from the crowds in the streets. Here and now. Counting the stairs. I used to do this when I was a kid. Seventeen steps to count. But sometimes the number changed, or seemed to. Did I miscount? Was the world shrinking or expanding? This was back then. People today tell me that they can't imagine me as a kid. Was I called Max? Growing up in a small town. Another thing they can't imagine. Mother, brother, sister. No angry crowds, no tall buildings. Seventeen steps. We were tenants, second floor of somebody's two-story house. Nine steps alongside the garage, then eight more to our apartment. A kid named Max. And suddenly here I am, a father, a man whose job takes him into luxury towers to inspect basements, stairways, rooftops, looking and finding violations of the building code. I love the violations. It justifies all my feelings about just about everything. Here and now, these crucial hours, I dodged and elbowed my way back to this street and this building and found my house key and unlocked the front door and did not have to remind myself, it's not even worth saying, that the elevators are not working, and I started slowly up the stairs, looking down at each step as I climbed, flight by flight, and I realized at some point that my hand was on the handrail and I decided that I didn't want it there and just climbed and counted, step by step, flight by flight. I'd like to say that I was reliving those earlier years but my mind was more or less blank. Just the stairs and the numbers, third floor, fourth floor, fifth floor, up and up and up, and then finally pushing through the door to the hallway and lifting the apartment key out from under the crumpled snotty handkerchief in my pocket, and now that I'm here I don't think I have to apologize for this long dumb description of climbing eight flights of stairs because the current situation tells us that there's nothing else to say except what comes into our heads, which none of us will remember anyway."

Tessa Berens studies the backs of her hands as if confirming the color, her color, and wondering why she is here and not somewhere else in the world, speaking French or a kind of splintered Haitian Creole.

"For years, many years, I've been writing in a little notebook. Ideas, memories, words, one notebook after another, a huge number by now stacked up in cabinets, desk drawers and elsewhere, and I revisit old notebooks sometimes and it amazes me to read what I thought was worth writing. The words carry me back into dead time. Little blue notebooks, maybe three inches by four inches, to be slipped into my jacket pocket, and I have dozens untouched at home, still to be filled. I take two or three whenever I travel and I look and listen and scribble something on a page. This is my journal. Means nothing to anyone but me. Could be a line of poetry that I will cross out seconds later. Could be an item on a supermarket shelf, the package design, the name of the product, take out the notebook, take out the ballpoint pen, so on and so on. But all I want to do now is go home. Jim and I. If we have to walk, fine, yes, in daylight. Will the sun be shining? Will the sun be in the sky at all? Who knows what any of this means? Is our normal experience simply being stilled? Are we witnessing a deviation in nature itself? A kind of virtual reality? This leads me to say that it's time to shut up, Tessa. When I say this, try to understand that it's not a self-critical comment but a point of self-importance. I write, I think, I advise, I stare into space. Is it natural at a time like this to be thinking and talking in philosophical terms as some of us have been doing? Or should we be practical? Food, shelter, friends, flush the toilet if we can? Tend to the simplest physical things. Touch, feel, bite, chew. The body has a mind of its own."

Martin Dekker up and down. Once again he stands and speaks, immersed in his nowhere stare.

“Time to stop, isn’t it? But I keep seeing the name. Einstein. Einstein’s Theory of Relativity causing riots in the streets or am I imagining this because it’s late and I haven’t slept and barely eaten and the people here with me are not listening to what I say. Einstein speaking beyond our current situation, which I’ve referred to as World War III. Einstein had no premonition concerning how this war would be fought but he made it clear that the next major conflict, World War IV, would be fought with sticks and stones. And the Special Theory, dated 1912, one hundred and ten years ago. Manuscript brown ink, unwatermarked paper and then the paper improves and the ink goes black. This is what I carry in my head, for better or bad or worse. What else? I need a shave. That’s what else. I need to look into a mirror and remind myself that it’s time for a shave. But if I leave this living room and walk into the lavatory, will I ever come back? Face in the mirror. Granular surveillance. Tech-dome. Two-factor verification. Gateway tracking. I can't help myself. The terms surround me. I try to think sometimes in a prehistoric context. A flagstone image, a cave drawing. All these grainy shreds of our long human memory. And then Einstein. The exhilarating language. German, English. 'Dependence of mass on energy.' I want to walk with him across the Princeton campus. Saying nothing, silent. Two men walking."

Then he says, "And the streets, these streets. I don't have to go to the window. Crowds dispersed. Streets empty."

This is what young Martin says, looking down into his parted fingers.

"The world is everything, the individual nothing. Do we all understand that?"

Max is not listening. He understands nothing. He sits in front of the TV set with his hands folded behind his neck, elbows jutting.

Then he stares into the blank screen.